B
C
D
A
1

Little People, **BIG DREAMS**®

CHARLES DARWIN

Written by
Maria Isabel Sánchez Vegara

Illustrated by
Mark Hoffmann

Frances Lincoln
Children's Books

Little Charles lived with his family in Shrewsbury, England, at a time when people didn't know much about the secret life of nature. Had worms always crawled on the ground? He wanted to find out!

Back then, many teachers told their students that animals and plants had appeared on Earth all at once, in the blink of an eye. Could this be possible? Somehow, Charles was not convinced…

Questioning things was natural for the Darwins. They were a family of scientists who thought outside the box. Charles' grandfather was a well-known plant expert and his father was a doctor who hoped his son might follow in his footsteps.

Charles entered medical school but he couldn't stand the sight of blood! So he switched subjects and continued to read and learn all about plants, animals, and nature.

One day, he received an invitation to join a scientific expedition to the coasts of Africa, South America, and Australia. The HMS *Beagle* left port during Christmas in 1831. It was a chance for Charles to see the world and finally study new species!

Soon, he started noticing how plants and animals changed from place to place. In Brazil, he found the remains of an animal that had lived thousands of years ago and was similar to other mammals he knew well.

On the Galápagos Islands, he met dozens of families of finches and mockingbirds. Those who ate insects had pointed beaks, while those who loved fruits had curved beaks. "What if these species shared the same ancestor?" Charles thought.

After five years traveling the world, Charles began to realize that plant and animal species were not fixed, as everyone thought. He formed the idea that they had slowly changed to adapt to the place in which they lived.

Charles was 28 years old when he penned one of the most shocking ideas of his time: that one species changes into another. It was the first step to put down on paper a revolutionary theory that explained how life on Earth works.

Over time, all living beings eventually become new species. This process happens through tiny little changes over thousands of years. We call it "evolution," and Charles had just discovered its secret mechanism.

Charles's discovery was called "natural selection:" nature rewards those that adapt best to their environment. The fastest rabbit, the smartest fox, or the owl with the best eyesight would survive in the wild and live to make many more animals like them.

It took Charles almost a lifetime to collect all his ideas in a book called *On the Origin of Species*. It was one of the most important books ever written and a fascinating read for anyone who wanted to understand the secret life of nature.

And since then, the latest discoveries in science are informed by Charles—the most important naturalist in history. The bold boy who understood that knowledge takes more courage than ignorance... but the truth is always worth it.

CHARLES DARWIN

(Born 1809 • Died 1882)

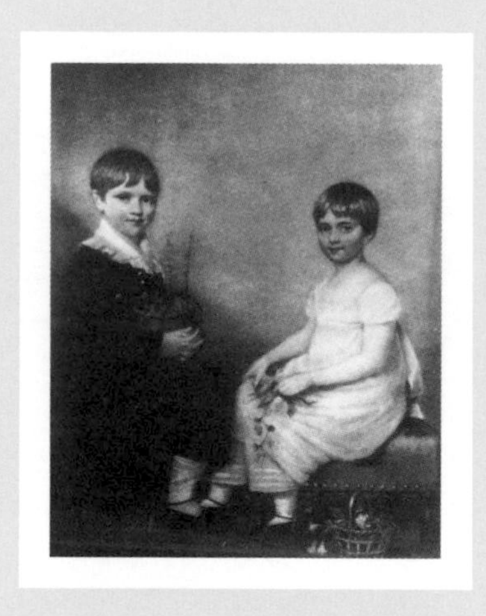

1816

1840

Born in the tiny merchant town of Shrewsbury, England, Charles Darwin arrived on February 12th, 1809. He quickly fell in love with nature as a young boy, taking long walks on his own to find nature specimens. When he was old enough, he enrolled in medical school but quickly dropped out to study theology at Cambridge. One of his theology professors suggested he pursue his interest in the natural world, recommending him as a companion for the HMS *Beagle* voyage. Unbeknown to Darwin, this voyage would dramatically change his life, as well as the future of scientific thinking. Aboard the *Beagle*, Darwin discovered a world he could only dream of: birds with bright blue feet, giant tortoises, and brightly-colored fish. He collected plants, animals, and fossils, and filled many notebooks

1854

c. 1878

with drawings and notes. During his time on the Galápagos Islands, he studied the beaks of finches. Some of these finches had stout beaks for eating seeds, while others were insect specialists. Darwin soon realized that they all shared a single ancestor—a revolutionary idea at the time. Collecting species from different islands, he discovered that each one had adapted to eat whatever food was available. Returning to England, Darwin knew his ideas would be met with opposition as they challenged religious ideas of his generation. After 20 years of research, he published the contentious but now-celebrated book, *On the Origin of Species*, and it quickly became a bestseller. Still today Darwin's work gives us insight into the diversity of life on Earth and its origins, including our own as a species.

Want to find out more about **Charles Darwin?**

Have a read of these great books:

When Darwin Sailed the Sea by David Long and Sam Kalda

On The Origin of Species by Sabina Radeva

Brimming with creative inspiration, how-to projects, and useful information to enrich your everyday life, Quarto Knows is a favorite destination for those pursuing their interests and passions. Visit our site and dig deeper with our books into your area of interest: Quarto Creates, Quarto Cooks, Quarto Homes, Quarto Lives, Quarto Drives, Quarto Explores, Quarto Gifts, or Quarto Kids.

Original concept of the series by Maria Isabel Sánchez Vegara, published by Alba Editorial, s.l.u

Produced under licence from Alba Editorial s.l.u and Beautifool Couple S.L.

First Published in the UK in 2021 by Frances Lincoln Children's Books, an imprint of The Quarto Group.

100 Cummings Center, Suite 265D, Beverly, MA 01915, USA.

T +1 978-282-9590 **www.QuartoKnows.com**

A catalogue record for this book is available from the British Library.

ISBN 978-0-7112-5771-9

Set in Futura BT.

Published by Katie Cotton • Designed by Karissa Santos

Edited by Katy Flint • Production by Nikki Ingram

Consultant: Dr Niall Sreenan, King's College London

Manufactured in Guangdong, China CC102020

1 3 5 7 9 8 6 4 2

Photographic acknowledgements (pages 28-29, from left to right): 1. 1816: British naturalist and author of 'The Origin of Species by Means of Natural Selection' Charles Darwin (1809–1882) and his sister, Catherine © Chalk drawing by Sharples; photo by Hulton Archive/Getty Images. 2. Portrait of a young Charles Darwin in 1840; watercolor and chalk on paper by George Richmond, 1840 © Photo by GraphicaArtis/Getty Images. 3. British naturalist Charles Darwin (1809–1882) who developed theory of evolution by natural selection © Photo by Time Life Pictures/Mansell/The LIFE Picture Collection via Getty Images. 4. Charles Darwin, (1809–1882), English scientist who developed the modern theory of evolution © Photo by Bob Thomas/Popperfoto via Getty Images/Getty Images.

Collect the
Little People, BIG DREAMS® series:

FRIDA KAHLO

ISBN: 978-1-84780-783-0

COCO CHANEL

ISBN: 978-1-84780-784-7

MAYA ANGELOU

ISBN: 978-1-84780-889-9

AMELIA EARHART

ISBN: 978-1-84780-888-2

AGATHA CHRISTIE

ISBN: 978-1-84780-960-5

MARIE CURIE

ISBN: 978-1-84780-962-9

ROSA PARKS

ISBN: 978-1-78603-018-4

AUDREY HEPBURN

ISBN: 978-1-78603-053-5

EMMELINE PANKHURST

ISBN: 978-1-78603-020-7

ELLA FITZGERALD

ISBN: 978-1-78603-087-0

ADA LOVELACE

ISBN: 978-1-78603-076-4

JANE AUSTEN

ISBN: 978-1-78603-120-4

GEORGIA O'KEEFFE

ISBN: 978-1-78603-122-8

HARRIET TUBMAN

ISBN: 978-1-78603-227-0

ANNE FRANK

ISBN: 978-1-78603-229-4

MOTHER TERESA

ISBN: 978-1-78603-230-0

JOSEPHINE BAKER

ISBN: 978-1-78603-228-7

L. M. MONTGOMERY

ISBN: 978-1-78603-233-1

JANE GOODALL

ISBN: 978-1-78603-231-7

SIMONE DE BEAUVOIR

ISBN: 978-1-78603-232-4

MUHAMMAD ALI

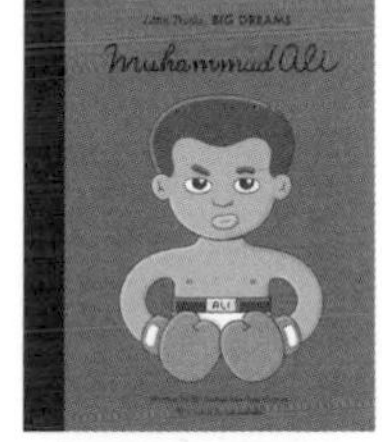

ISBN: 978-1-78603-331-4

STEPHEN HAWKING

ISBN: 978-1-78603-333-8

MARIA MONTESSORI

ISBN: 978-1-78603-755-8

VIVIENNE WESTWOOD

ISBN: 978-1-78603-757-2

MAHATMA GANDHI

ISBN: 978-1-78603-787-9

DAVID BOWIE

ISBN: 978-1-78603-332-1

WILMA RUDOLPH

ISBN: 978-1-78603-751-0

DOLLY PARTON

ISBN: 978-1-78603-760-2

BRUCE LEE

ISBN: 978-1-78603-789-3

RUDOLF NUREYEV

ISBN: 978-1-78603-791-6

ZAHA HADID

ISBN: 978-1-78603-745-9

MARY SHELLEY

ISBN: 978-1-78603-748-0

MARTIN LUTHER KING JR.

ISBN: 978-0-7112-4567-9

DAVID ATTENBOROUGH

ISBN: 978-0-7112-4564-8

ASTRID LINDGREN

ISBN: 978-0-7112-5217-2

EVONNE GOOLAGONG

ISBN: 978-0-7112-4586-0

BOB DYLAN

ISBN: 978-0-7112-4675-1

ALAN TURING

ISBN: 978-0-7112-4678-2

BILLIE JEAN KING

ISBN: 978-0-7112-4693-5

GRETA THUNBERG

ISBN: 978-0-7112-5645-3

JESSE OWENS

ISBN: 978-0-7112-4583-9

JEAN-MICHEL BASQUIAT

ISBN: 978-0-7112-4580-8

ARETHA FRANKLIN

ISBN: 978-0-7112-4686-7

CORAZON AQUINO

ISBN: 978-0-7112-4684-3

PELÉ

ISBN: 978-0-7112-4573-0

ERNEST SHACKLETON

ISBN: 978-0-7112-4571-6

STEVE JOBS

ISBN: 978-0-7112-4577-8

AYRTON SENNA

ISBN: 978-0-7112-4672-0

LOUISE BOURGEOIS

ISBN: 978-0-7112-4690-4

ELTON JOHN

ISBN: 978-0-7112-5840-2

JOHN LENNON

ISBN: 978-0-7112-5767-2

PRINCE

ISBN: 978-0-7112-5439-8

CHARLES DARWIN

ISBN: 978-0-7112-5771-9

CAPTAIN TOM MOORE

ISBN: 978-0-7112-6209-6

HANS CHRISTIAN ANDERSEN

ISBN: 978-0-7112-5934-8

ACTIVITY BOOKS

STICKER ACTIVITY BOOK

ISBN: 978-0-7112-6012-2

COLORING BOOK

ISBN: 978-0-7112-6136-5

LITTLE ME, BIG DREAMS JOURNAL

ISBN: 978-0-7112-4889-2

1
B
D
B
C
D
A
1